Made in United States
North Haven, CT
05 November 2026

21765618R00057